THE CHANGING OF THE GUARD

A full description of the Changing of the
Guard and other ceremonies

Un résumé en français est inclus

Auf Deutsch kurz zusammengefaßt

D0106759

LUTTERWORTH PRESS   LONDON

THE QUEEN'S GUARDS

Mounting the guard at Buckingham Palace and at the Horse Guards is one of the great ceremonies of the world, performed daily before eager crowds who have cause to admire the precision of the drill movements, the splash of colour of the scarlet uniforms, and the pageantry of the ancient ceremony.

Those who perform this ceremony are the troops of the Household Division, who, numbering 8,000, are the Sovereign's personal bodyguard. They comprise two mounted regiments, who provide the guard at the Horse Guards, Whitehall, and five regiments of Foot Guards who mount guard at Buckingham Palace, St. James's Palace, the Bank of England, and the Tower of London. The Household Division also provides guards and escorts for State occasions, and at the ceremony to mark the Queen's official birthday the battalion forming the Queen's Guard on that day 'troops its Colour'.

But although the tourist is inclined to think of these troops as being exclusively engaged in ceremonial activities, in fact half the Household Division is on service overseas at any one time. The Guards have been and are Britain's premier fighting soldiers — equipped with the most modern weapons, the discipline demanded by the precision of their drill, the thoroughness of their training, and the *esprit de corps* of their army life ensures they are formidable in battle. Their Colours proudly display their honours in war from the defence of Tangier in 1680 to the Battle of El Alamein in the Second World War, from the Napoleonic Wars to the defeat of Hitler.

(*Above*) The Guards in their winter coats.
(*Top left*) Mounted trooper of the Life Guards in winter dress.
(*Left*) Massed regiments of the Guards at the Trooping the Colour

3

THE GUARDS' DIVISION

The regiments which guard Buckingham Palace and St. James's Palace, as well as the Bank of England, and the Tower of London, are the regiments of Foot Guards.

Originally the older regiments had their own distinctive and colourful uniform, but in the mid-19th century they adopted the familiar style of scarlet tunics, dark blue trousers with a red stripe, and the bearskin head-dress. But though superficially the same, all regiments retain distinctive differences.

The photograph on the right shows five guardsmen in full-dress uniform. From left to right:

The Grenadier Guards are descended from the personal Guard raised by King Charles II in exile, known as the King's Regiment of Guards, and in 1697 as the First Regiment of Foot Guards. To commemorate their gallantry at the Battle of Waterloo (1815) they were renamed the First or Grenadier Regiment of Foot Guards.

A Grenadier's dress uniform can be recognised by the eight evenly spaced buttons and the white plume on the left side of the bearskin. Their badge is the royal cypher surrounded by a garter, while their collar badge is a grenade.

The Welsh Guards were raised in 1915, when King George V decided that the regiment should be formed so that all the countries in the United Kingdom would be represented by their own regiment.

The Welsh Guards wear their buttons in two groups of five and have a green and white plume on the left. Their badge shows a leek, the national emblem of Wales.

The Scots Guards were formed in 1642 by Charles II. During the Civil War the regiment was scattered, but was reformed in 1660 as part of the Scottish Army. In 1686 it became part of the English Army, known as the Third Regiment of Foot Guards. In 1831 the regiment was renamed the Scots Fusiliers Guards and finally became the Scots Guards in 1877.

The Scots Guards wear their tunic buttons in three groups of three, and have no plume in their caps. Their badge shows a thistle, the emblem of Scotland.

GRENADIER GUARDS WELSH GUARDS SCOTS GUARDS IRISH GUARDS COLDSTREAM GUARDS

The Irish Guards were formed by Queen Victoria in 1900 in recognition of the fighting qualities displayed by the Irish regiments in the Boer War.

The Irish Guards wear their tunic buttons in two groups of four, and have a blue plume in their bearskins. Their badge depicts a shamrock.

The Coldstream Guards are the oldest regiment – in terms of continuous service – in the British Army and possibly in the world. They are descended from 'Monck's Regiment of Foot'. After Cromwell's death in 1658, the regiment marched to London from their headquarters in the small Scottish town of Coldstream. Paraded on Tower Hill, King Charles II ordered them to lay down their arms and to take them up again as the Second Regiment of Foot Guards. It is said that not a man moved and that General Monck declared 'Sire, that regiment refuses to be known as second to any in the British Army.' His Majesty then ordered, 'Coldstream Guards, take up your arms', and to this day the regiment has the motto *Nulli secundus* (Second to none).

A Coldstreamer's tunic buttons are worn in pairs, and a red plume adorns his bearskin. The badge is the star of the Most Noble Order of the Garter – the highest order of chivalry in the land.

THE CHANGING OF THE GUARD

The Queen has a number of homes, both official (Buckingham Palace, St. James's Palace, Windsor Castle and Holyroodhouse in Scotland) and private (Sandringham and Balmoral), but it is only at the London and Windsor palaces that a daily guard is mounted.

The Queen's Guard in London is changed in the Forecourt of Buckingham Palace every day, save in May when the ceremony is mounted on some days in Horse Guards Parade. The time-table is shown on the next page. The Guard comprises two detachments, one each for Buckingham Palace and St. James's Palace, under the command of the Captain of the Queen's Guard.

When the Queen is in residence, which is indicated by the Royal Standard flying over the Palace, the detachments consist of the Captain, a Subaltern and an Ensign

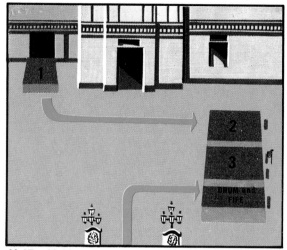

11·07 – 11·23 a.m.

1

2

3

DRUM and FIFE

PRIVY PURSE DOOR

6

5

NEW GUARD

BAND

4

11·30

10

9

OLD GUARD

NEW GUARD

11

–11·55

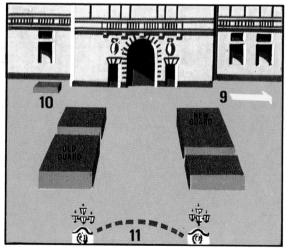

OLD GUARD

12

OLD GUARD

NEW GUARD

13

–12·05–

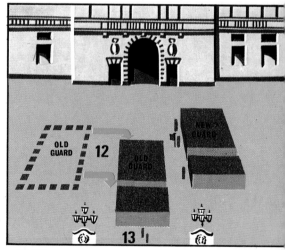

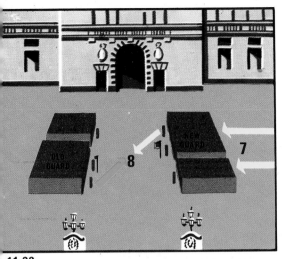

-11·33-

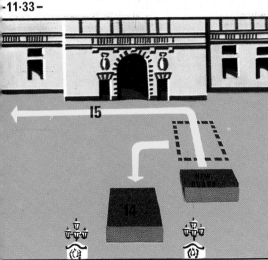

12·07-12·09

THE CHANGING OF THE GUARD

11.07   Buckingham Palace detachment parade outside Guardroom (1) and march to centre of Forecourt (2).

11.23   St. James's detachment enter Forecourt by South Centre Gate and form up on the right of the Buckingham Palace detachment to make the complete Old Guard (3).

11.30   New Guard enters Forecourt by North Centre Gate (4), halts and faces the Old Guard with the band behind them (5). Drill Sergeant dresses New Guard (6), which then advances in slow time (7).

11.33   The two captains hand over the Guard (8) and then enter the Palace by the Privy Purse Door to receive special orders (9). Ensigns, with colours, patrol at rear of Forecourt. Posting of relief sentries (10). The band plays a selection of music at Centre Gate (11).

12.05   Guards fall in as before. Old Guard advances in slow time (12) through Centre Gate, and break into quick time and march away (13).

12.07   St. James's detachment of New Guard marches out of Centre Gate (14) to St. James's Palace.

12.09   Buckingham Palace detachment marches to Guardroom (15).

(which is the name given to the rank of Second Lieutenant in the Foot Guards), and 40 guardsmen. If the Queen is not in residence the guard is reduced to three officers and 31 guardsmen.

At 11.07 a.m. the Old Guard at Buckingham Palace form up in the Forecourt and are inspected by the Subaltern (*left*). Meanwhile the St. James's Palace detachment has fallen in in Ambassadors' Court, has been inspected, and, led by the Corps of Drums, has marched off by the Stable Yard Gate to join the Buckingham Palace detachment. The photograph (*top, right*) shows this detachment entering by South Centre Gate of Buckingham Palace. At 11.23 a.m., as it enters, arms are presented by the Buckingham Palace detachment. The drums then form up on the right of the complete Guard (*right*), the Drum-Major being in line with the officers. The Drill Serjeant, who carries a pace-stick, ensures all guards are in their proper position and dressed in line. Throughout the ceremony the Colours never leave the Forecourt of the Palace.

While the Old Guard is preparing to dismount, the New Guard has formed up at its barracks together with the Tower of London Guard; it is inspected by the Adjutant and handed over to the Captain

of the New Guard. Accompanied by a regimental band of the Guards Division and by the battalion's own Corps of Drums (*left*), the Captain marches the New Guard to Buckingham Palace, arriving punctually at 11.30 a.m. and entering by the North Centre Gate.

When the New Guard has halted in line with the Old Guard, and the band and drums have taken up their positions, the Captain of the New Guard orders his guard to advance in slow time (*below*) while the band plays the appropriate slow march. Each Guards Regiment has its own quick and slow marches, for

Coldstream Guards respectively.

When the New Guard halts, the two guards present arms in turn (*left*).

The Captains of the two guards then bring their swords from the salute to the carry and advance towards each other to go through the motions (*below*) of handing over the Palace keys, though this is only symbolically performed.

instance the quick march of the Grenadier Guards is, not surprisingly, the ancient tune, *The British Grenadiers*, while the Irish Guards play the traditional air *St. Patrick's Day* as their quick march and *Let Erin Remember* as their slow march. The tune, *Milanollo*, is the quick march of both the Coldstream Guards and the Life Guards, while the latter have *Men of Harlech* as an alternative quick march and the Welsh Guards play it in slow time. The two marches of the Scots Guards are the Scottish tunes *Hielan Laddie* and *The Garb of Old Gaul*. Tunes from two operas, *Scipio* and *The Marriage of Figaro*, are the slow marches of the Grenadier and the

During the ceremony, the Captains of the Guards leave the parade and enter the Palace by the Privy Purse Door to receive any special orders for the day. Meanwhile other officers and N.C.O.s go to the Guardroom to sign over all effects.

The new sentries are posted. The complete orders are read to them (the orders seldom change), although they may have done sentry duties over a number of years. A sentry spends two hours on duty and four hours in the Guardroom.

THE COLOURS

The Standards of the Household Cavalry and Colours of the Foot Guards are the embodiment of a regiment's or battalion's historic traditions. In battle they indicated the whereabouts of headquarters and were paraded before the troops so that they would not fail to recognise them, however the Crimean War (1854–56) was the last time the Colours of the Foot Guards were carried into action. But even today a regiment needs to maintain its traditions and recall its deeds of valour, so the Colours, inscribed with a roll of battle honours, are treated with ceremony and respect.

The Regimental Colour of the Foot Guards has a background similar to a Union Jack. In addition, each regiment has a Queen's Colour, crimson in colour, which is carried by the Queen's Guard when the Court is in residence in London.

In this photograph, ensigns are parading the Colours of the Grenadier Guards. The wreath indicates that the regiment is commemorating a particular battle on that day.

(*Above*) The sentries of the Old Guard at St. James's Palace march into the Palace Forecourt to join the main ceremony.

(*Right*) The Drum-Majors and Pipe-Major return to the Forecourt.

(*Far right*) The Band-Sergeant shows a selection of music to the Captain of the Guard from which he selects the marches to be played on the return to barracks. All seven regiments of the Household Division have a band of highly proficient musicians. In addition, each of the Foot Guards battalions has a Corps of Drums and the battalions of Scots and Irish

Guards have pipers as well, the former wearing feather bonnets and tartan kilts and the latter green tam o'shanters and saffron kilts. At least two of these regimental bands are resident in London at any one time.

(*Below*) The ceremony at the Palace is almost over. The Old and New Guard stand at attention. The band, in this case that of the Scots Guards, is formed up at the Centre Gate. The whole parade now waits for the command to move off.

(*Overleaf*) The band of the Irish Guards returning to barracks. The Irish wolfhound, 'Fionn', is the only regimental mascot in the Household Division.

ST. JAMES'S PALACE

After Whitehall Palace was burnt down in 1698, St. James's Palace became the official residence of the Sovereign and ever since it has remained at the centre of the Court: for instance, foreign ambassadors are still accredited to the Court of St. James's. Besides, the Palace embraces Marlborough House, which was occupied by both Queen Alexandra and Queen Mary in their widowhood, and Clarence House, which is the London home of Queen Elizabeth the Queen Mother. As

St. James's Palace is still the official residence of the Court, it is here that the Colour is lodged and where also the Captain of the Guard establishes his headquarters.

(*Far left*) St. James's Palace detachment, led by the Pipes and Drums of the Scots Guards, coming away from Buckingham Palace to take up their duties at St. James's.

(*Left*) The Escort to the Colours march the Colours off to the officers' guardroom where they are lodged.

(*Above*) The Drum-Major seeks permission to return to barracks, the final act of the ceremony at St. James's.

THE QUEEN'S LIFEGUARD

When St. James's Palace became the official residence of the Sovereign in the 17th century the entrance was still considered to be at Whitehall, so it was natural that the Guard should be mounted at what came to be known as the Horse Guards archway. This Guard is provided by the two mounted regiments of the Household Division — The Life Guards, and The Blues and Royals (the Royal Horse Guards and 1st Dragoons). These two regiments are now fully equipped with tanks and armoured vehicles, but they retain a mounted squadron consisting of 128 horses each to provide the Sovereign's Escort and the Queen's Life Guard at Whitehall.

(*Left*) Dismounted sentry of The Life Guards, wearing winter cloak.

(*Above*) Trumpeters of The Life Guards, their trumpets magnificently adorned with trumpet banners bearing the royal coat of arms. Note their plumes are red compared with the white plumes of the rest of the squadron.

(*Right*) A mounted trooper of The Blues and Royals on sentry duty at the Horse Guards. Note his blue tunic and red plume compared with the scarlet tunic and white plume of The Life Guards. The sheepskin-covered saddles also differ, as those of The Life Guards are white, while those of The Blues and Royals are black. Both regiments wear cuirasses — making them the last British troops to wear armour.

Every morning the Queen's Life Guard rides from their new barracks at Hyde Park, down Constitution Hill, past Buckingham Palace and down the Mall to mount guard at the Horse Guards. If the Queen is in London the Guard consists of one officer, three non-commissioned officers and 11 troopers; a Standard is carried and the Guard is headed by a trumpeter on a grey horse. If the Court is out of London the Guard is reduced to two non-commissioned officers and 10 troopers; no Standard is carried.

(*Right*). Changing the Queen's Life Guard at Whitehall. The Life Guard are handing over to The Blues and Royals.

(*Above*). The Mounted Band of The Life Guards, led by the drum horses of both The Life Guards and The Blues and Royals.

24

THE QUEEN'S BIRTHDAY PARADE

On the day chosen as the Sovereign's official birthday in June, there takes place on the vast arena of the Horse Guards Parade a ceremony which has been called the greatest parade of all — Trooping the Colour in the presence of Her Majesty (*right*), the Colonel-in-Chief of all regiments of the Household Division. No other country presents such a spectacular military parade. The scarlet tunics, the waving plumes, the glistening breastplates of the Horse Guards, the perfect

grooming of the mounts, and, above all, the precision of the drill make it an unforgettable experience.

It is known that the Foot Guards carried out a Trooping the Colour parade as early as 1755, but it was in 1805 that the custom of performing the ceremony to honour the Sovereign's birthday was initiated. With a break from 1811 to 1820 (during the illness of George III), and also during the war years, it has been performed ever since. The regiments of the

Foot Guards take it in turns for their Colour to be trooped.

(*Above*) The massed Guards perform a complicated turning movement with the perfect precision for which the Foot Guards are renowned.

(*Left*) The drummer of The Life Guards riding 'Hector'. He is wearing State dress — velvet cap, gold coat, white breeches and boots. The drum, which is of solid silver, was presented to the regiment by William IV in 1831.

THE QUEEN'S KEYS

The regiment which provides the Queen's Guard at Buckingham Palace also finds the Guard for the Tower of London, which mounts on Tower Green at noon each day. Sentries are posted at the entrance and at the Jewel House, and every evening at 7 minutes to ten, the symbolic locking up of the Tower, known as the 'Ceremony of the Keys', starts. The public is not admitted, as the ceremony occurs when the Tower is closed, but it can be witnessed by a limited number on application to the Governor for a ticket. At the ceremony the Chief Yeoman Warder, carrying the Queen's Keys, is escorted by an armed guard (*top, left*) to the outer gate. On his return he is met by the Guard for the night, commanded by an officer with drawn sword. At the command 'Present Arms!', the Chief Warder takes two paces forward (*left*) and calls out 'God preserve Queen Elizabeth!' The whole Guard reply 'Amen!' and, as the clock strikes ten, the Last Post is sounded.

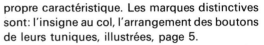

LA RELEVE DE LA GARDE

Ce livre décrit la cérémonie célèbre qui a lieu tous les jours à Buckingham Palace et à Horse Guards Parade à Whitehall. Il illustre aussi le grand apparat du « Trooping the Colour » (pages 26–28), qui s'accomplit chaque année en juin à l'occasion de l'anniversaire officiel de la Souveraine, et la « Cérémonie des Clefs » (page 29), qui a lieu quotidiennement le soir. Le grand public peut y assister en obtenant des billets du Gouverneur.

Ceux qui prennent part à ces cérémonies sont les Régiments de Gardes célèbres, comprenant: la Garde des Grenadiers et des Coldstreams, la Garde écossaise, irlandaise, galloise, la cavalerie Household de la Life Guard et les régiments des Bleus et des Royaux. Mais il faut se souvenir que ces régiments sont l'élite de l'Armée britannique, qu'ils sont munis des armes les plus modernes, étant en service actif dans plusieurs parties du monde. Les bataillons se relayent pour les fonctions de cérémonie bien que leurs fonctions principales soient le service de la garde du corps personnelle de la Souveraine.

Les uniformes des différents régiments de la garde à pied — tuniques écarlates, pantalons bleu foncé, chapeaux traditionnels de gardes, sont semblables, mais chaque régiment a sa propre caractéristique. Les marques distinctives sont: l'insigne au col, l'arrangement des boutons de leurs tuniques, illustrées, page 5.

On peut distinguer facilement entre les uniformes des deux régiments de la cavalerie Household; les Life Guards portent des tuniques écarlates, les régiments des Bleus et des Royaux ont des tuniques bleues.

Il y a deux cérémonies principales. La relève de la Garde à Buckingham Palace se passe dans l'avant-cour (sauf quelques jours en mai quand cela se passe à Horse Guards Parade). La relève commence le matin à 11.07 et dure une heure.

La série des photographies illustre le rassemblement de l'ancienne garde (page 10), l'arrivée du détachement de St. James's Palace (11), et celle de la nouvelle garde accompagnée de musique comprenant un corps de tambours (12). Ensuite sa progression dans l'avant-cour au pas ralenti (12), l'échange symbolique des clefs entre les capitaines des deux gardes (13), puis le départ des deux capitaines pour aller chercher toutes consignes spéciales (14) et le placement des sentinelles (14). Les photographies, page 20, montrent un détachement de la Garde qui défile vers St. James's Palace et dans Ambassador Court.

DIE WACHABLÖSUNG

Dieses Heft beschreibt die berühmte Zeremonie, die täglich vor dem Buckingham-Palast und in Horse Guards in Whitehall stattfindet. Es beschreibt auch das prunkvolle „Trooping the Colour" (Seiten 26–28), die Parade am offiziellen Geburtstag der Königin, und die „Schlüssel-Zeremonie" (Seite 29), die jeden Abend im Tower stattfindet. Hierfür kann beim Gouverneur des Tower eine begrenzte Zahl von Zuschauerkarten angefordert werden.

Die Soldaten, die diese Zeremonien ausführen, gehören zu den berühmten Garderegimentern. Dies sind fünf Infanterieregimenter, nämlich die Grenadier Guards, die Coldstream Guards sowie die schottischen, irischen und walisischen Guards, ferner zwei berittene Regimenter, die Life Guards und die Blues and Royals. Diese Regimenter sind Elite-Einheiten der britischen Armee; sie sind mit den modernsten Waffen ausgerüstet und werden überall eingesetzt wie jede andere Truppe. Es ist lediglich ihre oberste Pflicht, als persönliche Leibwache der Königin zu dienen. Bei den Zeremonien wechseln sie einander ab.

Die Galauniformen der Infanteriegarden — scharlachroter Waffenrock, dunkelblaue Hose und Bärenfellmütze — sehen auf den ersten Blick gleich aus. Sie unterscheiden sich aber bei den einzelnen Regimentern durch den Kragenspiegel und durch die Verteilung der Knöpfe am Waffenrock. Einzelheiten sind auf Seite 5 erklärt.

Die Uniformen der beiden Kavallerieregimenter sind leicht zu unterscheiden. Die Life Guards tragen scharlachrote Waffenröcke, die Blues and Royals blaue.

Die Wachablösung am Buckingham-Palast findet im Vorhof statt (nur an einigen Tagen im Mai in Horse Guards Parade). Die Zeremonie beginnt täglich um 11 Uhr vormittags und dauert eine Stunde.

Die Bildserie zeigt den Aufmarsch der abzulösenden Wache (Seite 10), die Ankunft der Abteilung vom St. James's-Palast (11), die Ankunft der neuen Wache mit Regimentskapelle und Trommlern, die im langsamen Schritt im Vorhof aufmarschiert (12), die symbolische Übergabe der Schlüssel zwischen den Hauptleuten der beiden Wachen (13), deren Abmarsch zur Entgegennahme besonderer Befehle und die Einweisung der Schildwachen (14). Die Bilder auf Seiten 20–21 zeigen eine Abteilung der Wache beim Marsch zum St. James's-Palast.

Die berittene Wache in Horse Guards zieht jeden Morgen um 11 Uhr auf (sonntags um 10 Uhr).

WHEN AND WHERE
TO SEE THE GUARDS

BUCKINGHAM PALACE. The Changing of the Guard ceremony commences at 11 a.m. daily.

ST. JAMES'S PALACE. St. James's detachment of the Queen's Guard parade at 11 a.m. before leaving for the main ceremony at Buckingham Palace, and return at 12.10 p.m. approximately.

Nearest Underground stations: Green Park; Trafalgar Square; or St. James's Park.

HORSE GUARDS, WHITEHALL. The Queen's Life Guard is mounted at 11 a.m. (Sundays 10 a.m.). It also parades dismounted at 4 p.m.

Nearest Underground stations: Trafalgar Square and Westminster.

THE TOWER OF LONDON. The Guard mounts at 12 noon daily. The Ceremony of the Keys takes place at 10 p.m.

Nearest Underground station: Tower Hill.

The Guards can also be seen out of London at **WINDSOR CASTLE,** Berkshire, which is another Royal residence. The Guard-mounting ceremony takes place daily at 10.30 a.m. in the Quadrangle when the Queen is officially in residence, and at other times outside the guardroom at the entrance to the Castle.

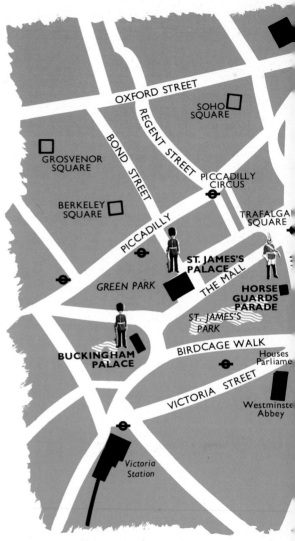